W9-BIA-925

LEWIS LEMON GLOBAL STUDIES
ACADEMY LIBRARY

LEWIS LEMON GLOBAL STUDIES
ACADEMY LIBRARY
Rockford, Illinois

Created and designed by
Aladdin Books Ltd
70 Old Compton Street
London W1

First published in the United States in 1989 by
Franklin Watts
387 Park Avenue South
New York NY 10016

Printed in Belgium

Design: David West
Children's Book Design
Editor: Roger Vlitos

Picture research: Cecilia Weston-Baker
Consultant: Dr. Marc Vlitos
Photographic credits:
Cover: Vanessa Bailey; pages 4-5, 25, 27 and 17tl: Robert Harding Library; pages 6-7, 13t and 15br: Spectrum Colour Library; pages 7, 11, 17tr, 18, 20, 21, 22t, 24, 27t and 30b: Roger Vlitos; pages and 29b: Zefa; pages 91, 10, 13b, 15m, 22b, 23 and 30t: J. Allan Cash Library; pages 9r, 21bl and br: Science Photo Library; page 14: Cowe/Network; page 15t: Sturrock/Network; page 15bl: Lewis/Network; page 17b: Goldwater/Network; pages 20t and 27b: Frank Spooner Agency; page 25: Rex Features; page 26: Reed/Magnum; pages 28 and 29t: The Metropolitan Police.

Library of Congress Cataloging-in-Publication Data

Twist, Clint
Facts On Alcohol/by Clint Twist.
p. cm.
Includes index
Summary: Examines what alcohol is, how it is made, why people drink it and become addicted to it, and how abuse of it can affect families.

ISBN 0-531-10821-X
1. Alcoholism-Juvenile literature. 2. Alcohol-Health aspects-Juvenile literature. (1. Alcohol. 2. Alcoholism.) I. Title. II. Title: Alcohol.
RC565.T845 1989
616.88'1--dc20 89-8896
CIP
Ac

Facts on
Alcohol

Clint Twist

FRANKLIN WATTS
New York · London · Toronto · Sydney

CONTENTS

MARTI
EXTRA
1971
APPELLATION ALS
CÔTES d
RIESI
LONDON DRY
GILBEY'S
DRAMBUIE
FREEZOMINT
CUSENIER
PARIS

INTRODUCTION

Alcohol is the most widely used drug known to man. It is drunk in all countries except where it is prohibited by religion and law. The manufacture and sale of alcohol is closely controlled by governments and everybody agrees that alcohol is a drug for adults only.

Alcohol can add to the enjoyment of life if it is used in moderation. Irresponsibe use of alcohol means trouble all the way – hangovers, accidents, health problems, personal problems, addiction and early death. Alcohol has a much greater effect on children and young people's health, because their bodies are not fully developed. Alcohol is also big business. Millions of people earn their living directly and indirectly from alcoholic drinks.

WHAT IS ALCOHOL?

When people talk about alcohol they nearly always mean the chemical known as ethyl alcohol or ethanol. Ethyl alcohol is the alcohol found in alcoholic drinks. There are other chemical forms of alcohol, but these are too dangerous to be used as a drink. When it is a pure chemical, the alcohol we drink is a colorless liquid with a strong smell and a fiery taste.

It is neither safe nor pleasant to drink pure alcohol. The chemical is far too strong for the human body to deal with. Our bodies can only cope with alcohol when it is mixed with water. Water, alcohol and different flavorings are the ingredients of almost all alcoholic drinks. The three main types are: beers, wines and hard liquors.

Alcohol is made from agricultural products. The many different types of alcoholic drinks come from different crops. In the United States and Europe, there is a very wide variety of alcoholic drinks available. As well as those produced locally, drinks are imported from all over the world. Different sorts of drink are usually served in different sized and shaped glasses. In some cases the size of the glass is the legal measure. The size and shape of the bottle also identifies certain drinks.

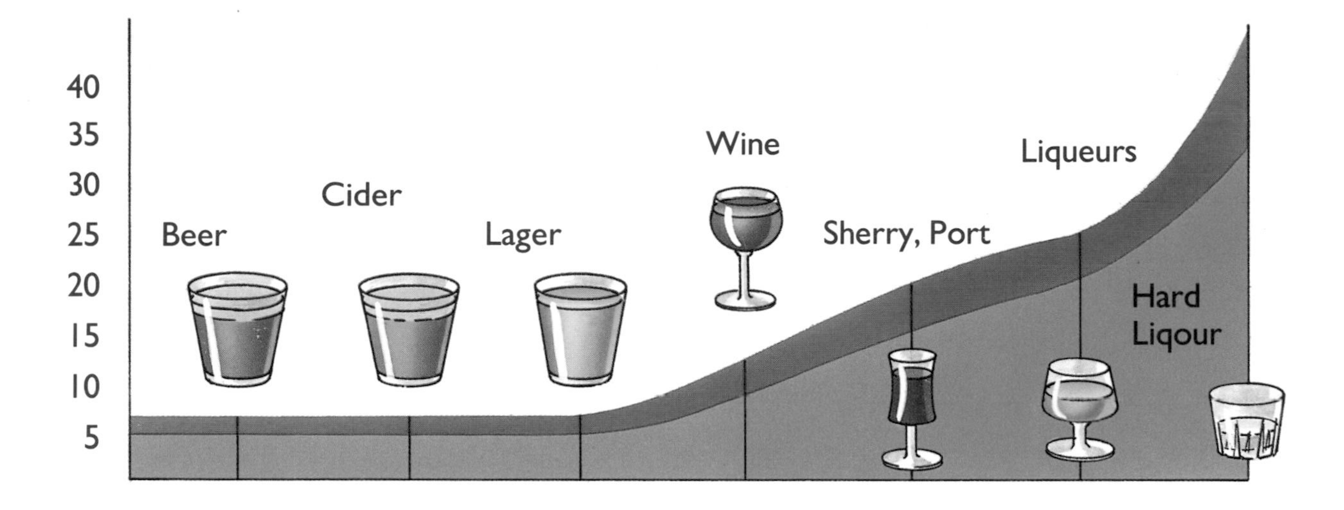

The alcoholic strength of drinks is expressed in different ways. The important thing is the percentage of alcohol by volume, which is called the "proof."

The amount of alcohol in drinks varies considerably. Beers contain only about 5 percent alcohol. Hard liquors are the strongest drinks with up to 50 percent alcohol.

OTHER ALCOHOLS

Chemical alcohols are used in the manufacture of a wide range of products. They include furniture polish, antifreeze for cars, perfumes and after-shave. These other alcohols are all extremely poisonous. Even small quantities of them can cause permanent harm if they are swallowed.

8 FERMENTING BEERS

Alcohol is made by the action of yeasts on sugar. Yeasts are microscopic organisms which belong to the same family as fungi. When they are mixed with sugar and water, yeasts convert the sugar and oxygen into alcohol and carbon dioxide gas. This process is called fermentation. Alcohol is really the yeast's waste product.

Beer making, or brewing, involves a number of stages. Fermentation is nearly the last of these stages. Beer is usually made from barley, although rice and other cereals can also be used instead. The other ingredients are water, yeast and hops (which gives beer its distinctive bitter flavor). Beer is made all over the world, and can have a very different color, flavor and alcohol content. It is exported double-strength from one country to another.

To make beer, barley grains are soaked in water, then drained and spread out while they are still damp. The grains start to sprout and produce a sugary substance called malt. The grains are then dried in a kiln, ground up and mixed with hot water in the "mash tun." Here the sugars in the malt are released. Hops are added to the mixture and it is boiled in a large "copper" or "kettle."

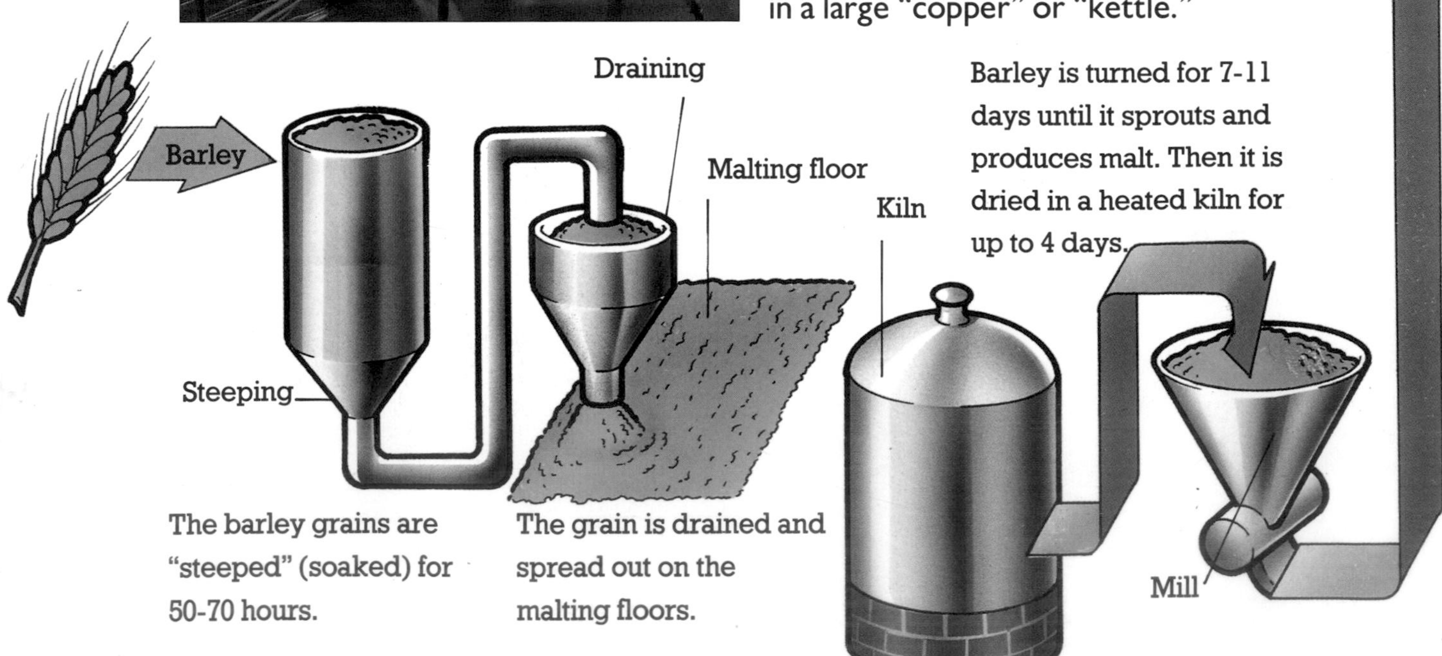

Barley
Draining
Malting floor
Kiln
Steeping
Mill
Barley is turned for 7-11 days until it sprouts and produces malt. Then it is dried in a heated kiln for up to 4 days.
The barley grains are "steeped" (soaked) for 50-70 hours.
The grain is drained and spread out on the malting floors.

DIFFERENT BREWS

The most common form of beer is called lager. It is made and sold in many countries around the world. It is pale in color, slightly fizzy, and is generally served cold. Traditional English beer, or "real ale," has no fizz and is served from wooden barrels.

Making a good beer requires precise control of ingredients, temperature and timing at all stages of the brewing process. Variations in the process produce different types of beer. Hops are not essential for brewing. They merely add flavor to the beer. Most of the yeasts used today have been specially cultivated for beer-making.

After the mixture has been boiled, yeast is added and fermentation begins. When it has been fermented, the beer is ready to be put into bottles, barrels or cans so that it can be sold.

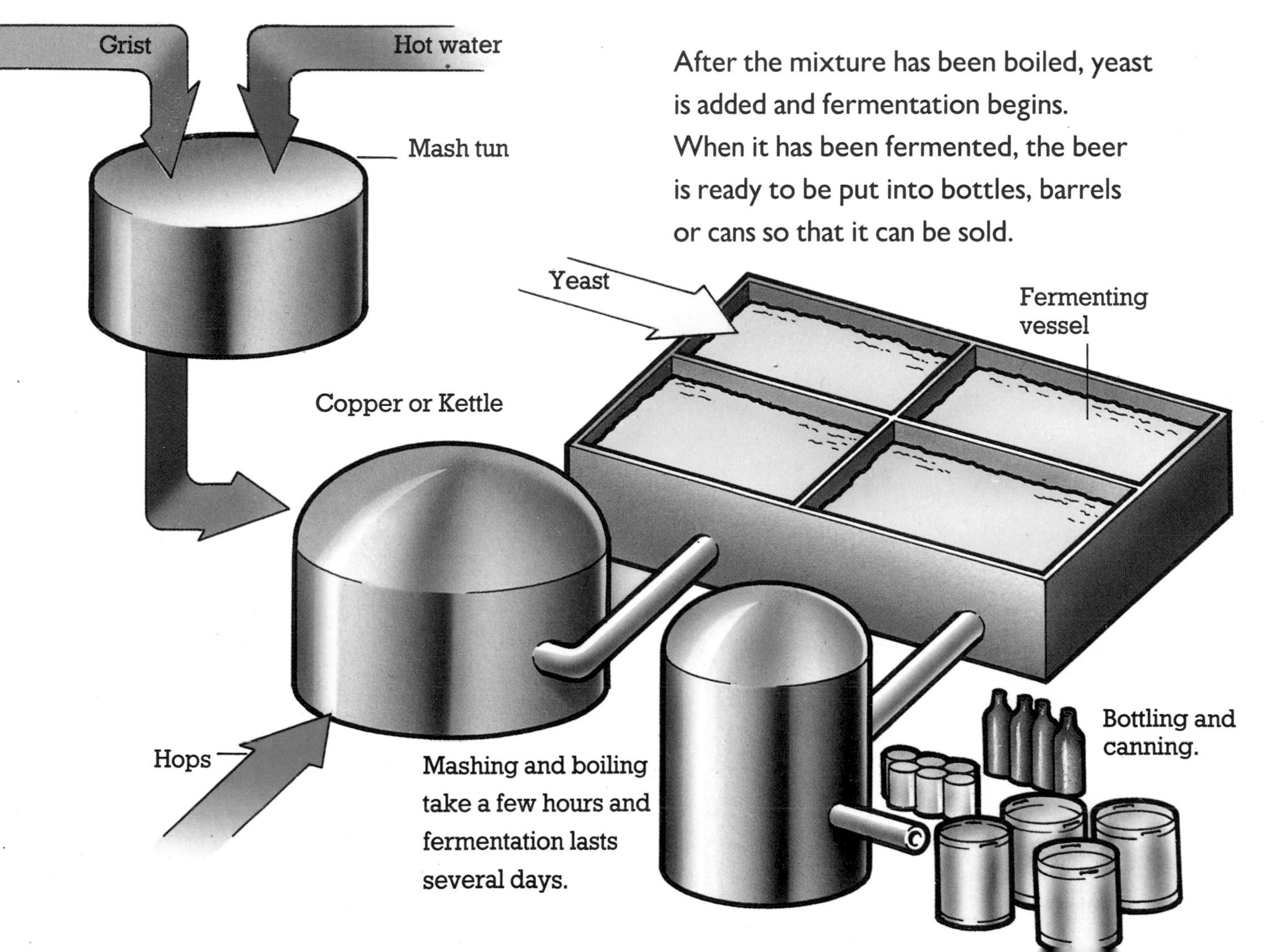

Mashing and boiling take a few hours and fermentation lasts several days.

FERMENTING WINES

Wine is made from the juice of grapes. This juice is rich in sugar, and the grape skins are covered with natural yeast. Fresh grape juice will ferment all by itself, but it must be protected from the air and bacteria, or it will turn into vinegar.

Most wine is now made and blended in modern wineries. But wine made from the grapes of a single vineyard is considered to have the best flavor. Wine is normally either red, white (pale yellow), or rosé (pink and slightly fizzy).

The main wine producers in Europe are France, Italy, Spain and West Germany. Most of the wineries in the United States are in California. Australia, Bulgaria, North Africa and Chile also export wine, and many other countries produce it for local consumption.

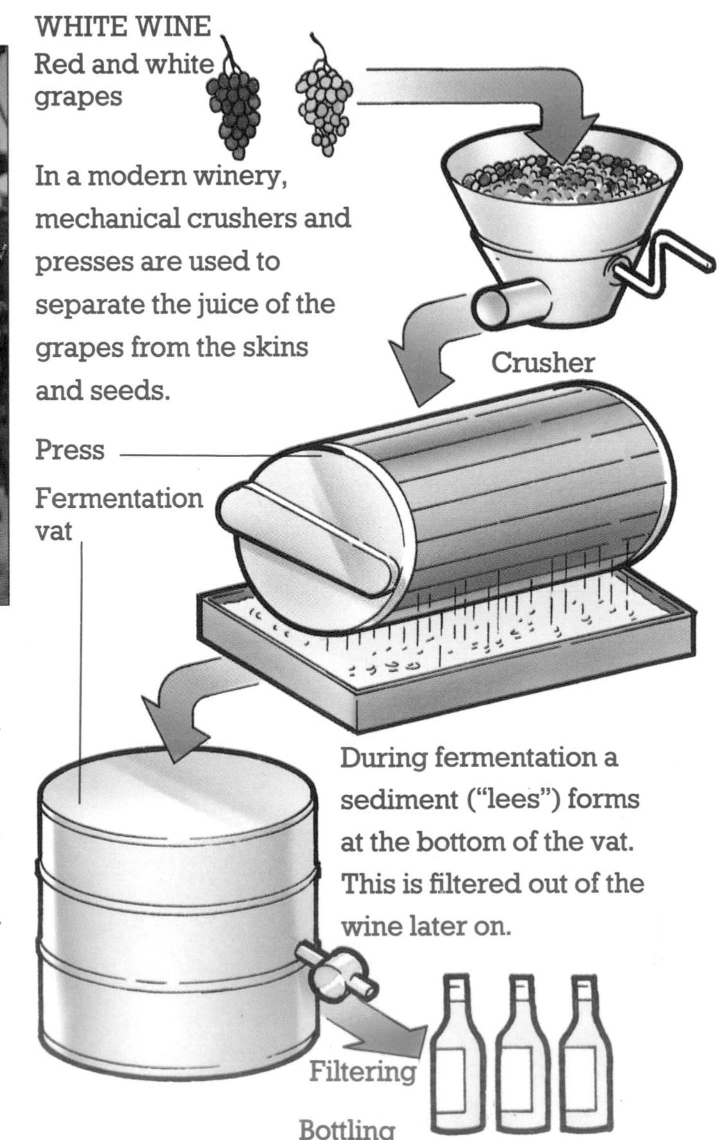

WHITE WINE
Red and white grapes
In a modern winery, mechanical crushers and presses are used to separate the juice of the grapes from the skins and seeds.
Crusher
Press
Fermentation vat
During fermentation a sediment ("lees") forms at the bottom of the vat. This is filtered out of the wine later on.
Filtering
Bottling

Inexpensive types of wine are sold as "table wine" or "vin ordinaire." More expensive wines are named after the region where they are made or the type of grape used. Different kinds of grapes produce distinctive flavors. The weather during the growing season is also important. Some wines are kept for many years to "mature" or improve their flavor.

A FRENCH WAY OF LIFE

Wine is a part of the everyday life of the French people. A great many adults drink it at both lunch and dinner. Children are often given wine mixed with water at an early age, and they grow up drinking wine like their parents. Perhaps, it is not surprising that France has more cases of liver disease than any other country in Europe. It also has a great many alcoholics.

White wine can be made from red grapes as long as the skins are removed after crushing. To make rosé wines, the skins are left a little while, and in red wine they are not removed until late in the fermentation process.

Most of the fermentation is over in a few days, but the wine remains in the vat for some time. It is then bottled and stored before being sold. Sparkling wines undergo a second fermentation inside the bottle. This produces the bubbles. Drinks such as sherry, port, and vermouth are known as fortified wines because they have been strengthened with extra alcohol.

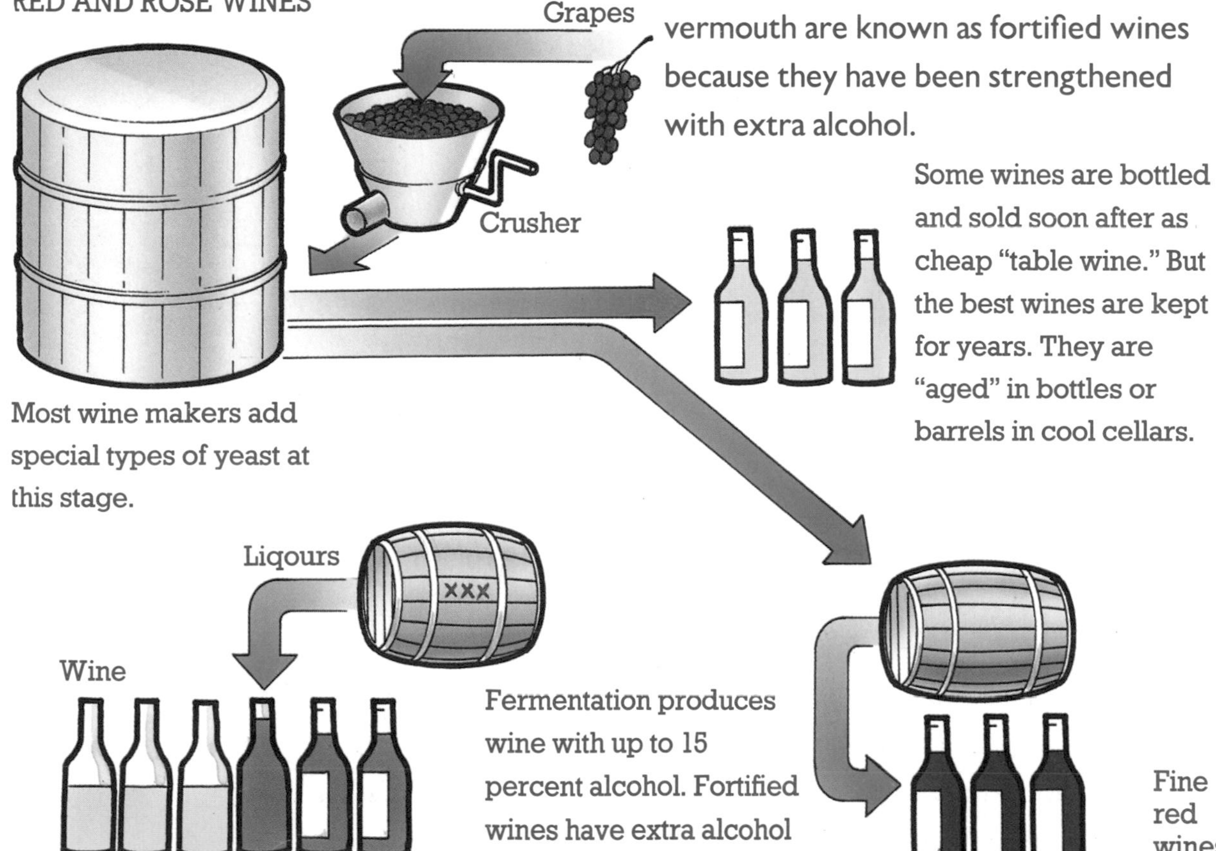

RED AND ROSÉ WINES
Grapes
Crusher
Most wine makers add special types of yeast at this stage.
Some wines are bottled and sold soon after as cheap "table wine." But the best wines are kept for years. They are "aged" in bottles or barrels in cool cellars.
Liqours
XXX
Wine
Fermentation produces wine with up to 15 percent alcohol. Fortified wines have extra alcohol added.
Fine red wines

DISTILLING HARD LIQUOR

Hard liquors or "Spirits" are the strongest alcoholic drinks. They are made by extracting the alcohol from a fermented liquid. This process is known as "distillation." It produces nearly pure alcohol. To make it safe to drink, this alcohol is mixed with water. Flavorings are also added for taste. On average hard liquors contain about 60 percent water.

Even so, hard liquors are so strong that they are served in very small quantities. They are often mixed with soft drinks or fruit juices. For example, gin is mixed with tonic water, rum with cola, vodka with orange juice. Sometimes two types are mixed together to make a drink called a "cocktail."

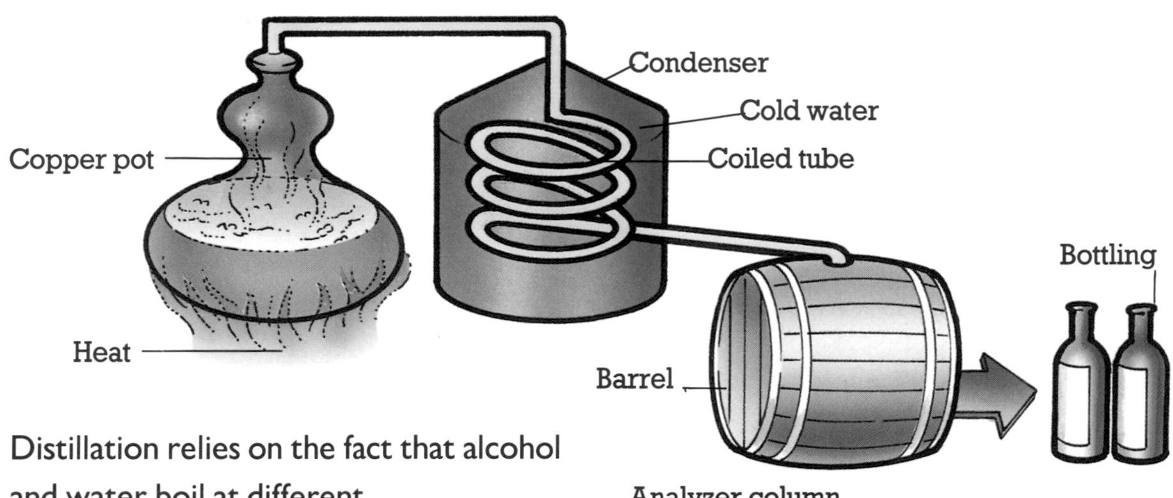

Distillation relies on the fact that alcohol and water boil at different temperatures. When the fermented liquid is heated, the alcohol boils before the water. The alcohol steam is then collected and cooled so that it finally condenses back to a liquid.

The equipment used in distilling is known as a still. Traditional stills are used in parts of Scotland and France. But most hard liquor is now produced in modern distilleries which use two hollow columns to condense the alcohol on metal plates inside the copper.

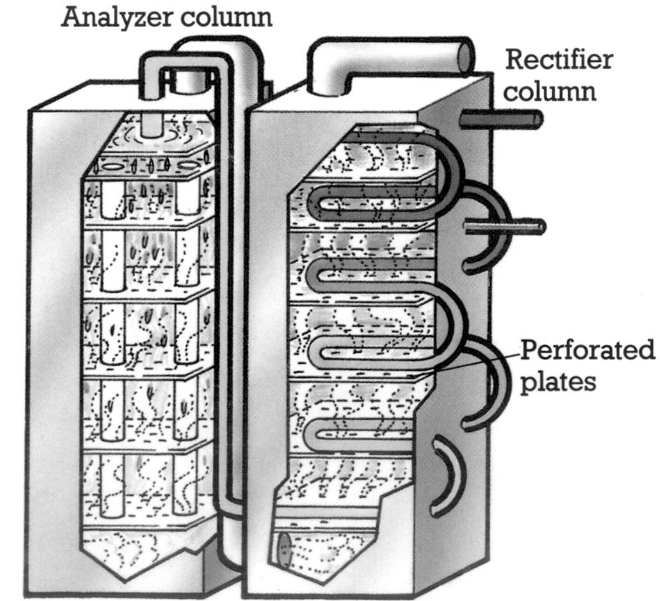

A wide variety of crops can be used to distill hard liquor. Brandy is distilled from wine, and whisky is made from a cereal mash like that used in beer-making. Most vodka is produced from potatoes. Rum is made from sugar-cane and tequila comes from cactus juice.

Many hard liquors have flavor added after they are distilled – gin gets its bitter taste from juniper berries. Liqueurs are highly flavored and contain a lot of added sugar. This hides the taste of the alcohol and can make them seem less strong than they really are.

PROHIBITION

Many countries allow people to make wine and beer at home, but distilling is usually illegal. In some countries bootleg hard liquors or "moonshine," are made in rural areas. These often contain impurities which poison the people who drink them.

Between 1919 and 1933 the United States government banned the sale of all alcoholic drinks. These "Prohibition" laws, as they were called, did not stop people drinking. Instead, they helped bootleggers to make millions of dollars.

WHY DO PEOPLE DRINK ?

The use of alcohol is deeply embedded in the social life of many countries. Alcohol is drunk at major events, such as births, weddings and funerals. Offering drinks to visitors often forms part of everyday hospitality. Alcohol is often associated with celebrations, relaxation and good times.

Beer quenches the thirst and wine adds to the enjoyment of a good meal. But perhaps the biggest reason for adults drinking alcohol is for the effect it has on the mind. They like the way it makes them feel light-headed, "tipsy," or "happy."

Making and selling alcoholic drinks is a very profitable business all over the world. Companies spend millions of dollars each year persuading us to drink their particular brand of alcohol.

Many young people drink alcohol simply because their friends do. They feel that they have to drink in order to be part of the group. They also feel that it makes them appear older or more fashionable.

Making a sensible decision is much more important than following the crowd. Alcohol is a powerful drug and people should not drink it unless they really want to. Part of the reason that people get addicted to alcohol is forgetting they have a choice over what they do.

ALCOHOL IN BUSINESS

Drinking alcohol forms a key part of business hospitality. Deals are often discussed at a "business lunch" where people tend to drink and talk a lot. Without thinking, they often drink more than is good for them. Business people may feel they have to drink in order to keep up with the others. This is the same sort of pressure that is felt by younger drinkers. They end up drinking because they feel they have to.

Advertisements often make alcohol seem exotic and very glamorous. Sometimes they seem humorous. However, they always mislead us into thinking that alcohol is a harmless, light-hearted drug.

16 THE PRESSURE TO DRINK

A glass of wine, a measure of hard liquor, or a small glass of beer, all contain about the same amount of alcohol – what doctors call a "standard unit." A pint of beer and a double whisky each contain two units. Most drinkers are young adults (aged 18 to 34 years). In this age group, about seven people out of every ten drink regularly. The number falls to just under six out of ten people in the over 35-age group. What gives most cause for concern is that nearly a third of all young people aged 12 to 17 also drink alcohol regularly, even though in most countries it is strictly illegal for them to buy alcoholic drinks.

Users of alcohol in age groups **Users of alcohol in %**

12-17 years 18-25 years 26-34 years 35 & over

Young people like to drink in groups. A few drinks can make everyone in the group seem more friendly. Stupid drinking games are sometimes played and the group gets over-excited and noisy. This sort of behavior often leads to complaints from other members of the public. Holidaymakers have a bad reputation for drinking too much in countries where alcohol is cheap.

When the group has another purpose, such as supporting a football team, the use of alcohol can have more serious results. Over-excitement can quickly turn to violence against rival supporters. The whole group, even those who are not normally aggressive, can become involved in fighting. If it seems that violence could develop, the police can confiscate alcohol from fans, or prohibit its sale. This can also be a problem at rock concerts.

It is clear, therefore that alcohol affects all age groups in society. Its use, or misuse is the responsibility of everyone.

Some of the worst drinking goes on at parties. Everybody wants to be friendly and they often drink too much and too quickly. Very easily, people can find themselves drunk and being sick, instead of enjoying themselves.

The dangers of alcohol are now recognized and there are a wide variety of alcohol-free, or low-alcohol drinks available. These taste very similar to beer or wine but they contain either little alcohol or none at all.

WHAT ARE THE EFFECTS ?

Alcohol affects both the mind and the body. The effects on the mind last a matter of hours: the effects on the body can be permanent.

Alcohol acts as a "depressant" drug on the brain. It is called a depressant because it slows down our reactions. Alcohol is carried to the brain in the blood. Here it gradually affects us, changing our moods and the ability to talk, walk and think if we drink too much.

Alcohol acts as a long-term poison on the body. Small amounts of this drug can be processed by the human body without any lasting harm. But over a long period of time, too much alcohol can lead to serious illnesses and perhaps even death. Health problems are common among heavy drinkers.

LOOSENING UP

The first effect of alcohol is usually to make people more lively. What actually happens is that after one or two drinks people start to lose self-control. They may feel less nervous and more talkative. Alcohol seems to stimulate them, when in fact it is putting their self-consciousness to sleep, and depressing the brain.

The depressant effects of alcohol become more obvious after a few drinks. Your vision becomes blurred and reflexes become much slower. Skills, such as driving, are also badly affected. People who have been drinking often have difficulty walking properly.

Once in the stomach, alcohol is absorbed straight into the bloodstream and carried round the body. It starts acting on the brain very quickly and continues to work for a long time whilst it is in the blood.

The liver slowly removes alcohol from the blood and breaks it down. This process produces water and carbon dioxide gas. The carbon dioxide is let out of the body through the lungs. The water passes out of the body in urine.

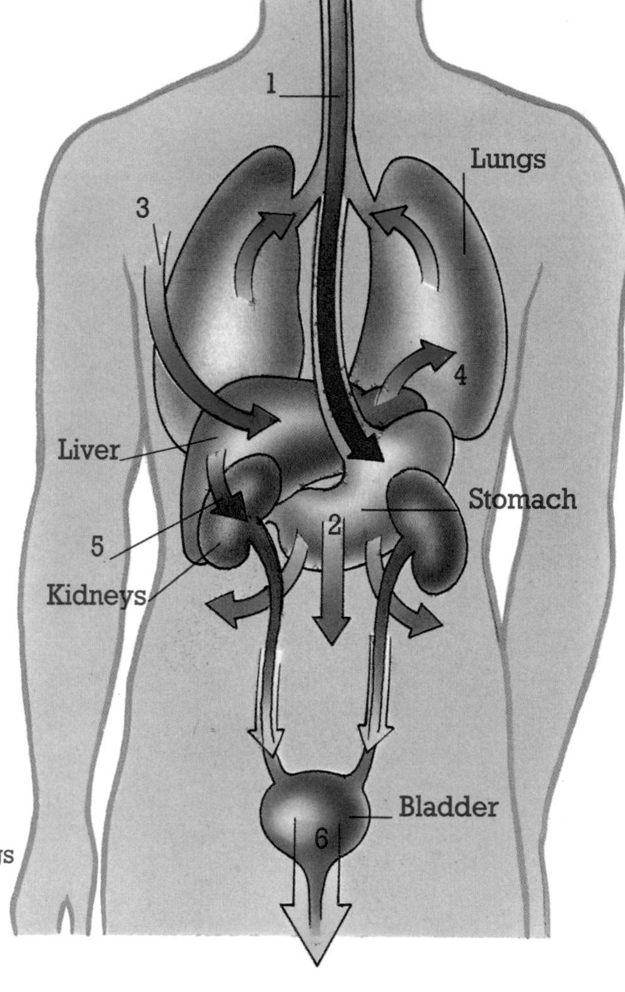
1
Lungs
3
4
Liver
Stomach
2
5
Kidneys
Bladder
6

1.Alcohol enters the stomach
2.Alcohol enters the bloodstream
3.Bloodstream back to the liver
4.Carbon dioxide goes from the liver to the lungs
5.Waste water goes to the kidneys
6.Urine is discharged from the bladder

The liver takes about an hour to process one unit of alcohol. People usually drink faster than this so the level of alcohol in the blood can build up dangerously.

The higher the level of alcohol in the bloodstream, the greater the effects and the longer the liver takes to process, or remove, alcohol from the body.

Processing time

A pint of beer	2hrs
A glass of wine	1-2hrs
A unit of hard liquor	1-2hrs

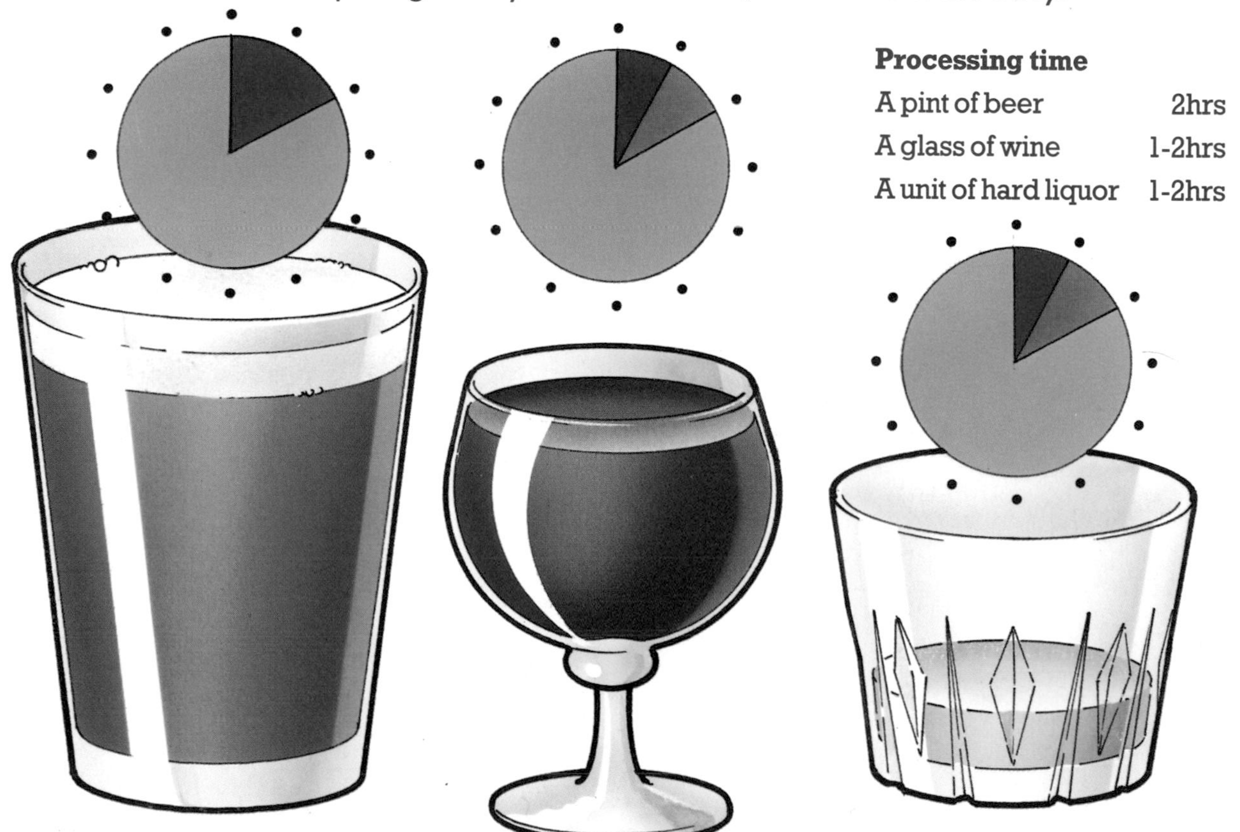

UNDER THE INFLUENCE

WHAT HAPPENS ?

People seem stupid when they are drunk. They get confused, and find it difficult to speak. Sometimes they get angry and violent. Often they feel dizzy and want to be sick. Excessive drinking leads to a hangover the next day which is very unpleasant.

THE HANGOVER

A hangover is a reminder that alcohol is a poison. It causes a bad headache and stomach pain by dehydrating the body. There is no quick cure for a hangover.

The effects of alcohol depend on many factors. Generally men can drink more than women before getting drunk, and somebody with a heavy build can drink more than somebody with a light build. People who drink regularly build up a tolerance to alcohol. They do not feel the same effect as an occassional drinker.

2 UNITS
Accidents become more likely.

3 UNITS
Cheerful and relaxed. Affected judgement.

5 UNITS
Illegal to drive. Accident risk 4 times higher than for 2 units.

10 UNITS
25 times accident risk of 2 units. Quarrelsome, excitable, slurred speech.

12 UNITS
Memory-loss, Staggering, Double-vision.

25 UNITS
Black-out and collapse.

32 UNITS
Death possible.

LONG-TERM EFFECTS

Some of the brain cells put to sleep by alcohol do not wake up – they die. This damage to the brain is permanent and can affect the memory.

Prolonged use of alcohol can cause painful and dangerous ulcers which eat through the walls of your stomach.

Alcohol has also been linked to high blood pressure and heart disease, several forms of cancer, and serious damage to other internal organs.

Depression, memory loss
Throat and stomach cancer
High blood pressure
Hepatitis and liver cirrhosis
Stomach ulcers and inflammation
Damage to pancreas
Vitamin deficiency
Nerve damage

Women suffer more than men. Abuse of alcohol reduces fertility. Drinking during pregnancy is known to increase the risk of a deformed baby.

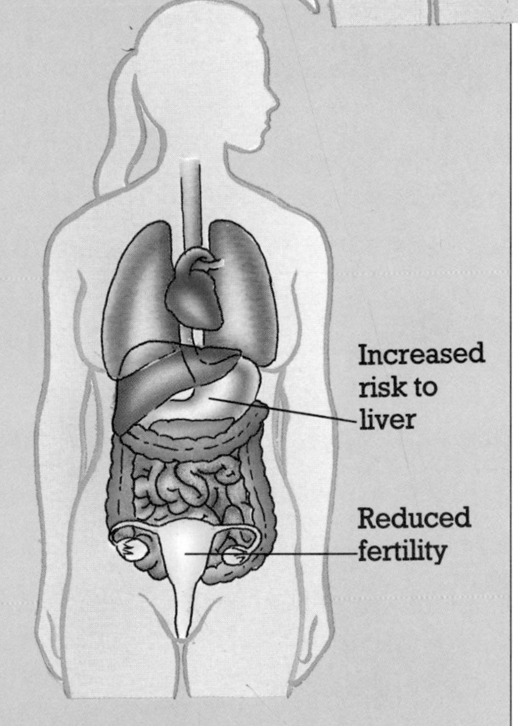

Our liver keeps the blood free of poisons. Alcohol abuse can cause an awful disease called cirrhosis. This condition occurs when the liver is overloaded with work removing alcohol from the body. Its cells gradually harden and shrink. Poisons build up in the body and the sufferer becomes seriously ill and can die.

A healthy liver

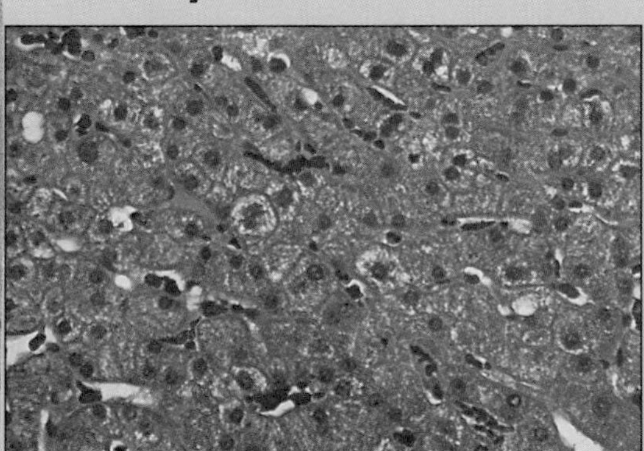

A diseased liver

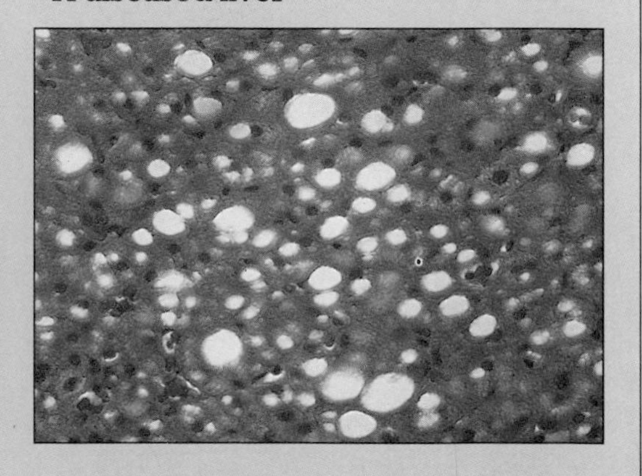

ADDICTION

Alcohol is a habit-forming drug. Some people who drink alcohol want to go on and on drinking even when they know it is harming them. We say they have become addicted. Whatever the reason for their excessive drinking, such people are known as alcoholics. Very few of them will admit it. In spite of what we might suppose, most alcoholics seldom get drunk. They take small, regular drinks to keep up the level of alcohol in their body. But their body chemistry changes so that they feel they need alcohol in order to get on with everyday life. In addition to increased health risks, alcoholics face other problems. Work and family relationships may be badly affected by their habit. Breaking alcohol addiction is not easy because both the brain and the body crave a drink.

Anyone can become an alcoholic. There is no special type of person, or age limit. Everyone is at risk when drinking. The truth is that alcohol is a drug.

Many alcoholics try to conceal their problem. Bottles of alcohol are hidden away at home or at work. Some even manage to live apparently normal lives until they get physically ill. Others lose all control over their own lives and end up jobless, friendless, and miserable because of their drinking.

DRINKING HABITS

Women and men have very different drinking habits. Women often weigh less than men, and that means they have less water in their bodies to dilute the alcohol they drink.

As a result, women tend to feel the effects of alcohol faster than men do. Perhaps this is why ten times more men become heavy drinkers. It might also explain why twice as many women than men only drink occassionally.

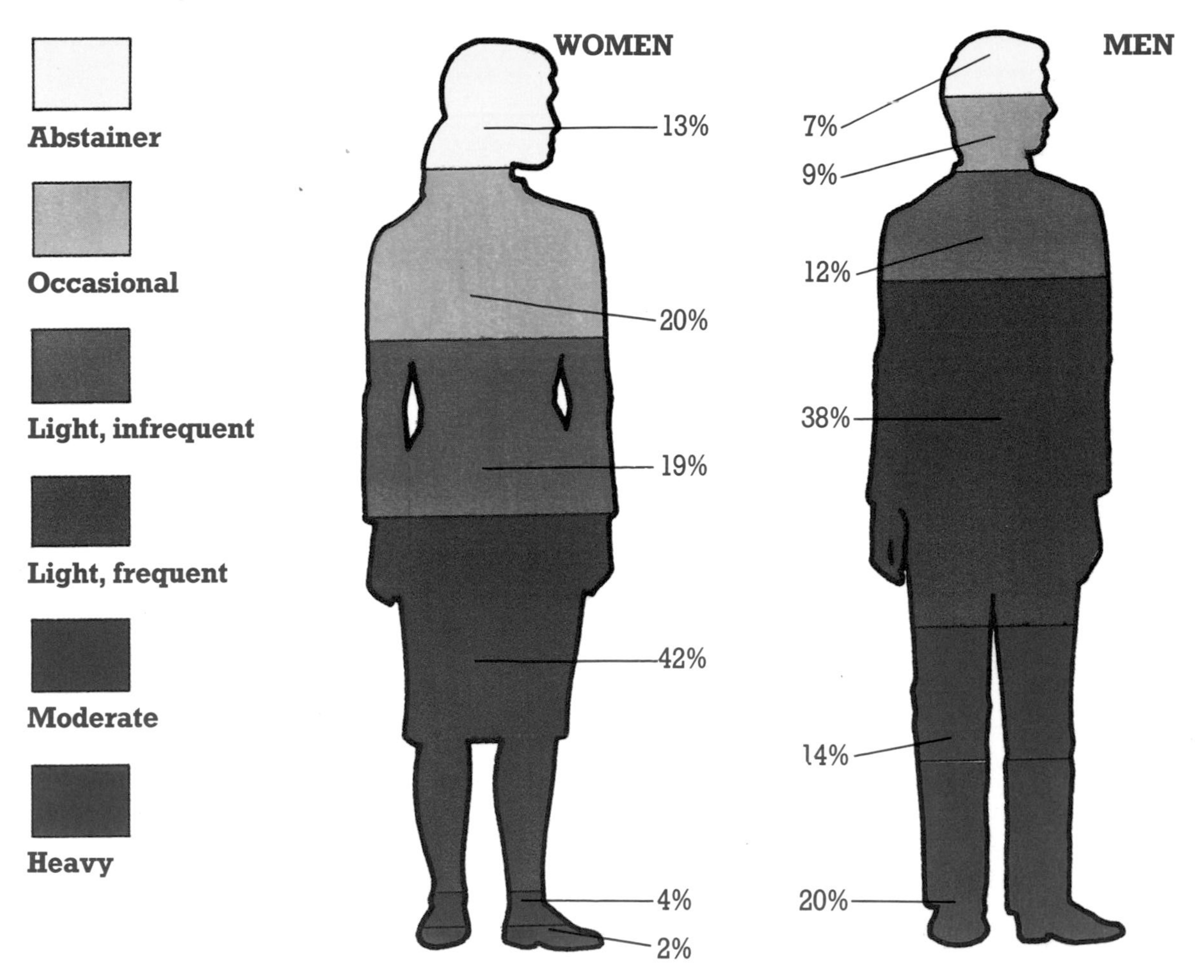

A WORLD PROBLEM

Alcohol addiction – alcoholism – is a growing problem. So much so, that in some countries alcoholism is seen as a common illness. Alcoholism affects far more men than women. Some people never drink at all – about 13 percent of women and seven percent of men. As many as one woman in fifty and one man in five is a heavy drinker. Their drinking could develop into alcoholism.

COPING WITH ADDICTION

A DRINK PROBLEM

People who are depressed can try to drown their sorrows in drink. Doing this regularly can make a person dependent on alcohol. People with a drinking problem often drink by themselves; and when they drink with friends they always finish their drink first. After a while, they need a drink first thing in the morning, just to "get them going." Such people have become alcoholics even if they do not know it.

Drinking alone

Finishing before friends

Coping with problems

Physical dependence

HELP FROM FRIENDS

Alcohol can also exaggerate a person's worries, making them seem worse than they really are. People who have been drinking sometimes get very unhappy. The alcohol confuses them and they lose sight of all the good things in life.

Even if they are talking nonsense, such people need gentle reassurance and calming down. Sometimes talking with friends can help a person face up to his or her worries and decide how to deal with them.

THE "DTs"

Alcoholics find giving up alcohol very difficult and unpleasant. Without a regular drink they start to panic and feel sick. Their hands may tremble (the shakes), and they will probably have bouts of sweating and shivering.
In a few cases there is a period of temporary madness and loss of memory. Some people see imaginary animals, others feel insects crawling under their skin. These hallucinations are commonly called the "DTs" *(delirium tremens)*.

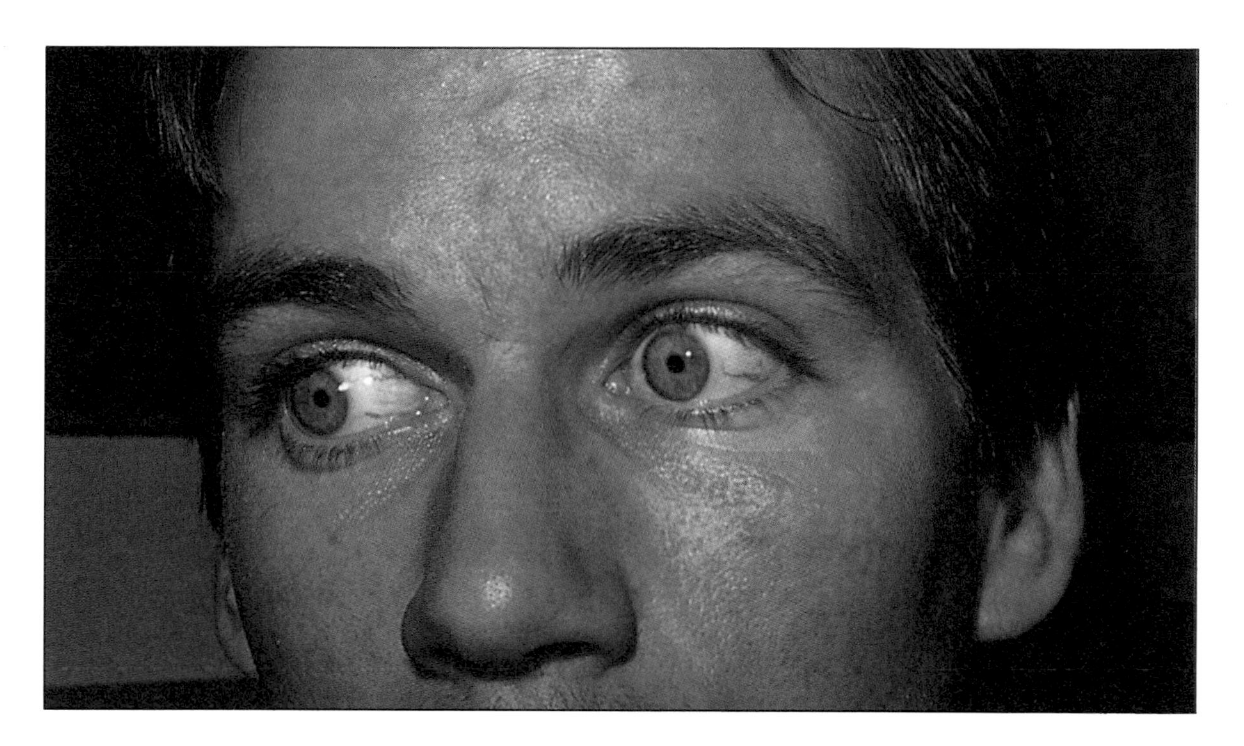

There are organisations which support alcoholics who want to give up drink. Members of a group called Alcoholics Anonymous help each other do this.

Some alcoholics benefit from group therapy. Understanding their reasons for drinking often helps them to stay away from alcohol and rebuild their lives.

ALCOHOL IN SOCIETY

Millions of gallons of alcoholic drinks are consumed daily, bringing pleasure to millions. However, alcohol also brings pain, misery and death to thousands of people every day. Alcohol is too popular to be completely banned – Prohibition did not work in the United States and similar laws are broken in strict Moslem countries.

Most countries limit the availability of alcohol in one way or another. A minimum drinking age is usually strictly enforced, as are rules against drinking and driving.

People are more likely to argue, fight, or commit crimes after they have been drinking. The things they stop doing are also important – home life and children are often neglected by people with a drink problem.

Some parents find it difficult to break old drinking habits. Others drink to get away from noisy children. They end up losing their tempers and hit their children. When a parent drinks it leads to arguments and wasting money. Families break up because of alcohol. Some children have to be taken away from their parents.

ACCIDENTS AT WORK

Because alcohol affects our judgement, accidents are more common when people drink. Captains of ships, airline pilots, and machinists in factories are all forbidden drink on the job. Tragedies, such as the oil-spillage shown on the right, can occur when they ignore this rule.

TURNING TO CRIME

Many crimes are committed by young people who have been drinking alcohol. After a few drinks they often feel more aggressive and confident. Crimes such as assault, burglary and stealing cars often occur after drinking sessions. Some people without the money to buy alcohol simply steal it. They may even take cash from their parents and friends. Others resort to mugging.

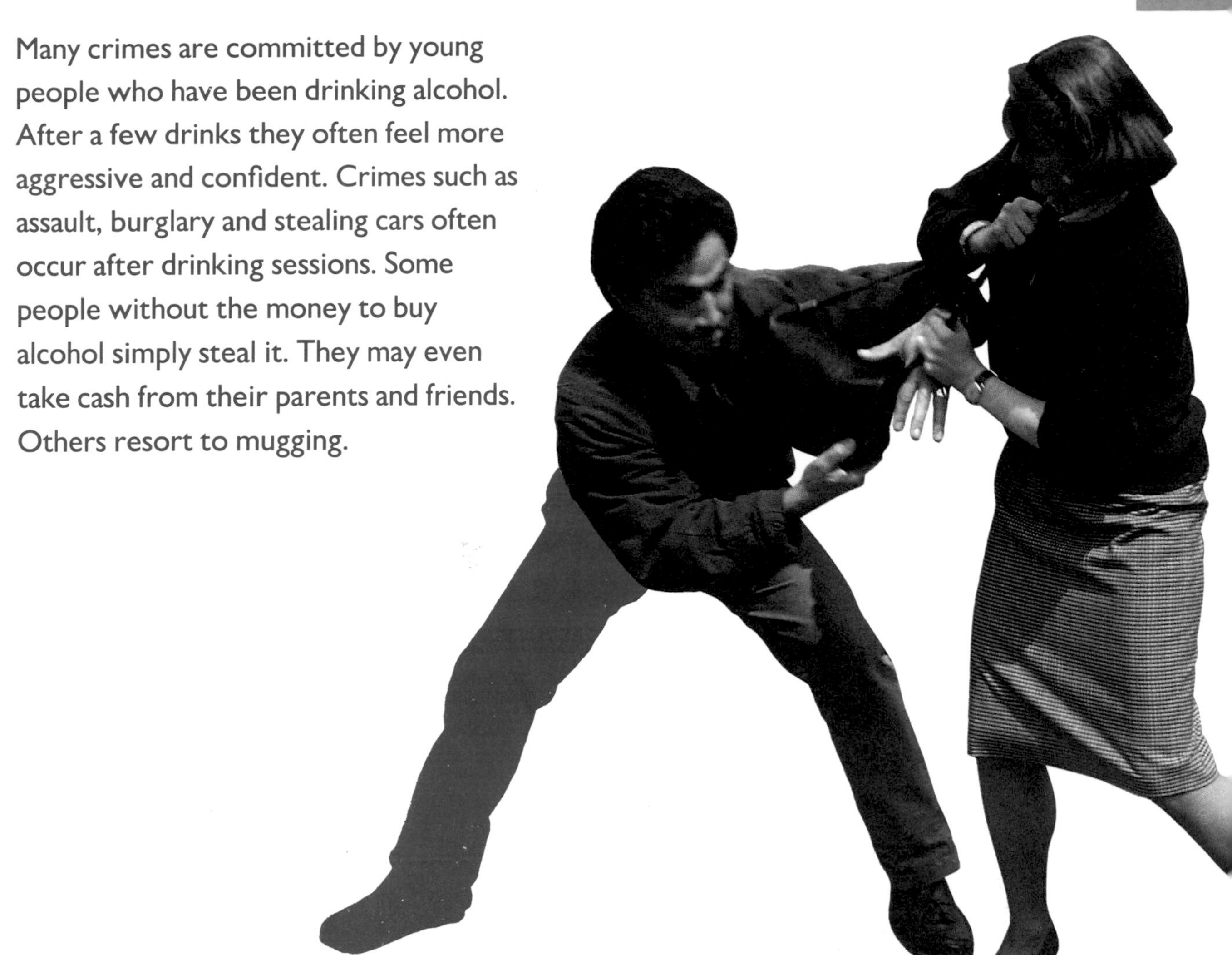

DRINKING AND DRIVING

In the United States, alcohol-related road accidents are the commonest cause of death in young men. People who drink and drive are not only being stupid with their own lives. They are selfishly endangering others. Passengers and pedestrians may also be killed or crippled for life. There is no excuse for drinking and driving. Government advertising hammers the message home – "save lives, don't drink and drive." With a little planning, everyone can avoid mixing alcohol and driving. Friends can share the cost of a taxi. Or they can take it in turns to have a night off alcohol and do the driving. They might even find that they enjoy themselves just as much.

It is against the law to drive with more than a certain level of alcohol in the blood. People who do are said to be "over the limit." The limit is very low, and even one drink can put you over.

The same rules apply to bicycling and swimming – "don't do it after drinking." Remember, alcohol increases the chance of accidents occuring.

THE BREATHALYZER

After you have had an alcoholic drink, there is alcohol on your breath, and in your blood and urine. In most countries it is illegal to drive a motor vehicle with more than a certain amount of alcohol in the bloodstream. Police often use a breathalyzer to measure the alcohol in the breath and to check if a driver is over the limit. If so, the driver is taken to a police station for a more accurate test to be made on a sample of urine or blood.

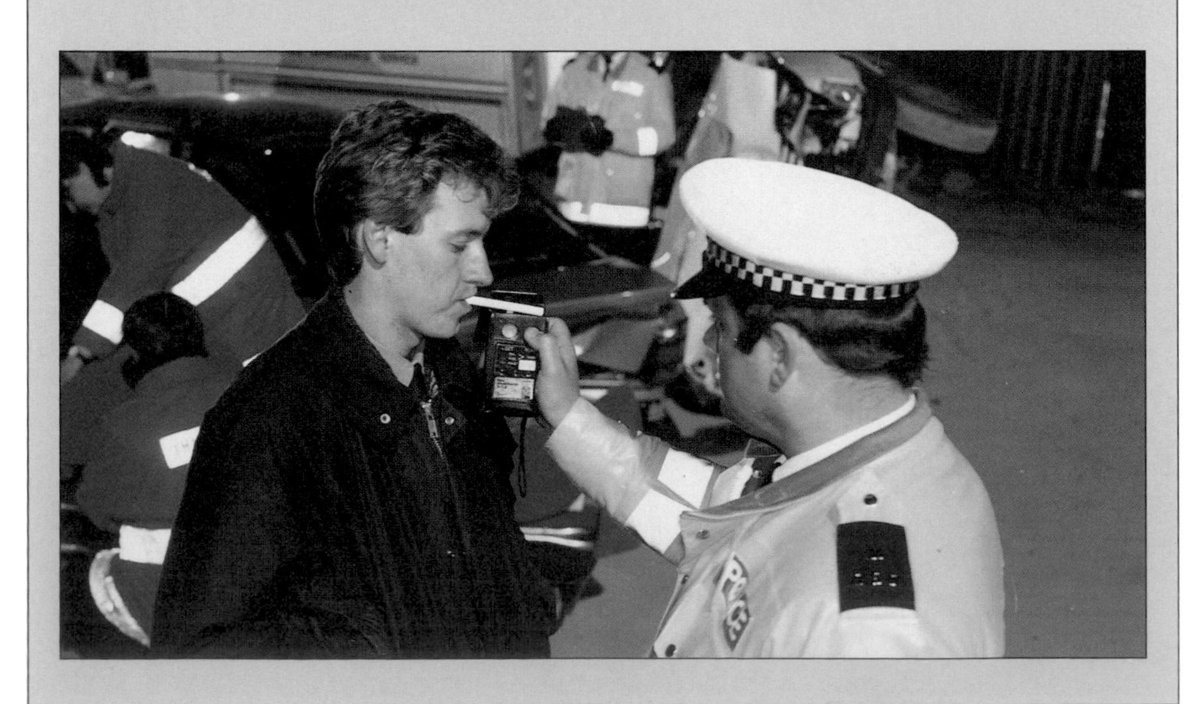

CONSEQUENCES

Being drunk is no excuse for committing a crime. In fact drunkenness itself is generally against the law. Courts have little sympathy with people who fight, steal or cause damage while under the influence of alcohol. Violent hooligans receive severe punishments. Under age drinking often leads to an unpleasant encounter with the police. Getting a criminal record is not a sensible way to start life. People who are caught drinking and driving usually lose their driving licence. This can badly affect people's lives. Some people could lose their jobs if this happened.

STATISTICS

Many automobile accidents are related to drinking, which is why there are laws about drinking and driving. Many offenders are young people. In the United States nearly one in every six drink and driving offences are committed by people aged between 18 and 20. In Britain, a third of all such offenders are under 25 years old.

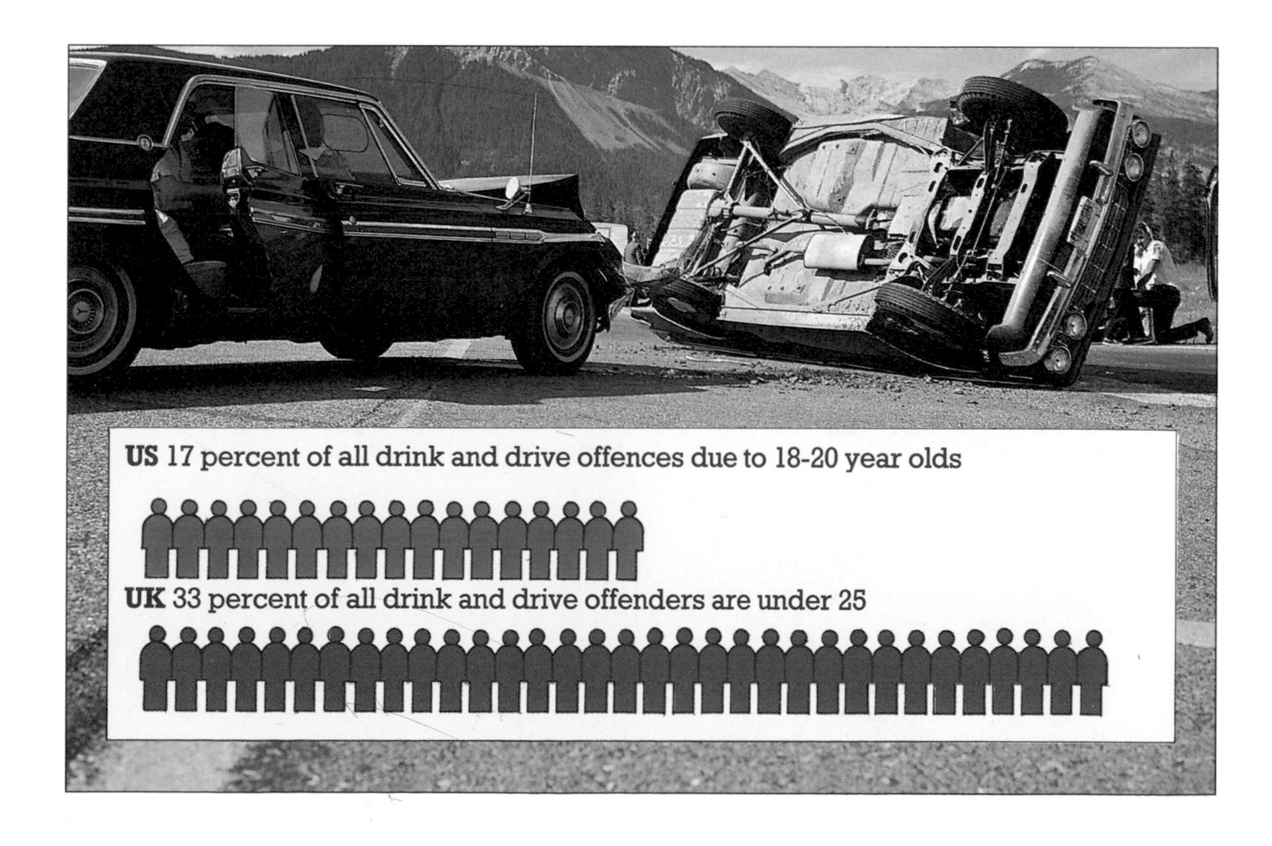
US 17 percent of all drink and drive offences due to 18-20 year olds
UK 33 percent of all drink and drive offenders are under 25

LEGISLATION

In the United States, the public use of alcohol is controlled by laws that specify exactly where and when alcohol can be sold. The laws are different in different states. In some states, the law prohibits the sale of alcohol to people under 18 years old. In other states, the age limit is 21 years of age. It is also against the law to buy alcohol for someone who is under 18 (21). The Under-18 (21) law does not apply if you are at home, but many parents find the law a useful guideline. Doctors also agree that alcohol can be particularly harmful to young people whose growth and development has not been completed.

To be able to sell alcohol for drinking on the premises, bars need a special licence. Stores which sell alcohol for drinking off the premises have another type of licence.

In general, people under 18 (minors) are not allowed into bars where alcohol is sold. If police catch a minor drinking in a bar that person may be arrested and the owners of the bar may also be punished. This may mean the licence to sell alcohol is taken away.

It is a serious offence to drive while intoxicated. The amount of alcohol a motorist is allowed to have in the blood before he or she is said to be legally intoxicated differs from state to state. At present, the police can only breathalyze motorists who they suspect have been drinking. Drinking and driving has become so much of a problem that many people want police to be allowed to test anybody.

It is against the law to be drunk in a public place. Although there is no legal definition of drunk (other than drunk while driving), the courts will accept the opinion of the police officer. It is also an offense to have unwrapped alcohol in a public place or to have open alcohol in an automobile.

GLOSSARY

Addiction – being mentally, and therefore physically, unable to do without something. People who are addicted to a drug are said to be unable to do without it.

Alcohol – by itself alcohol usually means the chemical called ethyl alcohol, or ethanol, the active ingredient of alcoholic drinks.

Alcoholic – a person addicted to alcohol.

Drink – any beverage which contains alcohol.

Drug – any chemical substance that is taken to affect the brain and body of a person.

Drunk – under the influence of alcohol to the extent that neither the brain nor the body are functioning properly.

Hangover – the aftereffects of drinking too much alcohol.

Liver – one of the body's principal organs. A strong liver is essential to purify the blood, and keep our bodies healthy.

Problem drinking – regular and excessive consumption of alcohol which can or will lead to alcoholism.

Units – an amount of alcohol equivalent to a glass of beer, a glass of wine, or a measure of hard liquor.

SOURCES OF HELP

Alcoholics Anonymous
PO Box 459
Grand Central Station
New York, NY 10017
(212) 696 1100
AA is a self-help organization, run by ex-alcoholics. You do not have to stop drinking before you can join. Local chapters are listed in the phone book.

Al-Anon Family Groups
PO Box 862
Midtown Station
New York, NY 10018
A support group for families and friends of people with drinking problems. Al-Teen is a section of Al-Anon specifically for young people.

National Institute of Drug Abuse
Treatment Referral
1-(800) 622-H-E-L-P

INDEX

PRINTED IN BELGIUM BY proost INTERNATIONAL BOOK PRODUCTION